T0225235

Cambridge Elements ☰

Elements in Geochemical Tracers in Earth System Science
edited by
Timothy Lyons
University of California
Alexandra Turchyn
University of Cambridge
Chris Reinhard
Georgia Institute of Technology

EARTH HISTORY OF OXYGEN AND THE IPROXY

Zunli Lu
Syracuse University, New York

Wanyi Lu
Syracuse University, New York

Rosalind E. M. Rickaby
University of Oxford

Ellen Thomas
Yale University and Wesleyan University

CAMBRIDGE
UNIVERSITY PRESS

CAMBRIDGE
UNIVERSITY PRESS

University Printing House, Cambridge CB2 8BS, United Kingdom

One Liberty Plaza, 20th Floor, New York, NY 10006, USA

477 Williamstown Road, Port Melbourne, VIC 3207, Australia

314–321, 3rd Floor, Plot 3, Splendor Forum, Jasola District Centre,
New Delhi – 110025, India

79 Anson Road, #06–04/06, Singapore 079906

Cambridge University Press is part of the University of Cambridge.

It furthers the University's mission by disseminating knowledge in the pursuit of
education, learning, and research at the highest international levels of excellence.

www.cambridge.org
Information on this title: www.cambridge.org/9781108723381
DOI: 10.1017/9781108688604

First published 2020

A catalogue record for this publication is available from the British Library.

ISBN 978-1-108-72338-1 Paperback
ISSN 2515-7027 (online)
ISSN 2515-6454 (print)

Earth History of Oxygen and the iprOxy

Elements in Geochemical Tracers in Earth System Science

DOI: 10.1017/9781108688604
First published online: September 2020

Zunli Lu
Syracuse University, New York

Wanyi Lu
Syracuse University, New York

Rosalind E. M. Rickaby
University of Oxford

Ellen Thomas
Yale University and Wesleyan University

Author for correspondence: Zunli Lu, zunlilu@syr.edu

Abstract: How oxygen levels in Earth's atmosphere and oceans evolved has always been a central question in Earth System Science. Researchers have developed numerous tracers to tackle this question, utilizing geochemical characteristics of different elements. Iodine incorporated in calcium carbonate (including biogenic) minerals, reported as I/Ca, is a proxy for dissolved oxygen in seawater. Here we review the rationale behind this proxy, its recent applications, and some potential future research directions.

Keywords: upper ocean, I/Ca, deoxygenation, carbonate, hypoxic

ISBNs: 9781108723381 (PB), 9781108688604 (OC)
ISSNs: 2515-7027 (online), 2515-6454 (print)

Contents

1 Introduction

The history of oxygen, as both a planetary biosignature of and a requisite for all complex life, has been the focus of a growing research community for decades (e.g., Holland, 2006). To understand and trace how the surface of Earth evolved from largely anoxic to an oxygenated system of rich ecological diversity, numerous proxies have been proposed, developed, and cross-examined thoroughly in a wide variety of geologic records (e.g., Lyons et al., 2014). More and more nuances in Earth's oxygen history are being revealed, revising earlier conceptual models into reconstructions of oxygen levels in four dimensions (3-D space and time) at ever increasing spatial and temporal resolution. Rather than the "classical" model of a two-step rise in oxygen to modern levels in Earth's atmosphere and oceans, additional oxygenation events have been proposed for the generally low-oxygen world of Proterozoic and even early Paleozoic times (e.g., Dahl et al., 2010; Diamond et al., 2018). In addition, an increasing number of oxygen depletion events have been suggested as punctuating the high-oxygen world of the Phanerozoic Eon (e.g., Jenkyns, 2010). There is an emerging need to better constrain and calibrate proxies for individual components of the Earth system (e.g., atmosphere, surface ocean, deep ocean), and for individual redox windows (e.g., oxic–hypoxic, hypoxic–anoxic, anoxic–euxinic). Regardless of the wave of paleo-redox proxy development sweeping across the periodic table, there are, clearly, open niches for proxies based on carbonate records, proxies specifically targeting relatively shallow and surficial parts of the water column, and proxies for the oxic–suboxic redox window. For such applications, the carbonate iodine-to-calcium ratio (I/Ca) has been developed as the iodine proxy for oxygen (iprOxy).

2 Proxy Systematics

Iodine is a redox-sensitive biophilic element. Its fluxes in/out of seawater are small relative to the amount of iodine recycled within the water column through biological production and remineralization (Lu et al., 2010). Iodine has a long residence time of ~300 kyr (Broecker and Peng, 1982), and it has a biointermediate element distribution rather than a nutrient-like or scavenging distribution (Farrenkopf et al., 1997b). The redox chemistry of iodine in the atmosphere and seawater may involve many different iodine species. However, the iodine species most relevant to the I/Ca proxy are iodide (I^-) and iodate (IO_3^-), which are thermodynamically most stable in anoxic and oxic seawaters, respectively. A number of studies have been dedicated to the redox chemistry of iodine in seawater (Amachi et al., 2007; Cutter et al., 2018; Farrenkopf et al., 1997a; Farrenkopf and Luther, 2002; Luther and Campbell, 1991; Luther et al.,

1995; Rue et al., 1997; Wong and Brewer, 1977; Wong and Cheng, 2008; Wong and Zhang, 2003; Wong et al., 1985). The reduction of iodate to iodide is generally fast, leading to the predominant presence of iodide in low-oxygen water, with very limited exceptions (e.g., Cutter et al., 2018; Rue et al., 1997). The oxidation of iodide to iodate can be kinetically slow (e.g., Hardisty et al., 2020), which may cause low concentrations of iodide to persist in oxygenated waters (e.g., Chance et al., 2014).

Laboratory crystal synthesis experiments (Lu et al., 2010; Zhou et al., 2014), synchrotron studies (Kerisit et al., 2018; Podder et al., 2017), and modeling (Feng and Redfern, 2018) show that iodate is incorporated in the calcite structure by substituting for the carbonate ion with charge compensation through Ca^{2+} substitution (e.g., Na^+), whereas iodide may be excluded from the mineral lattice. Hence, I/Ca values of carbonate should record changes in iodate concentrations in the water. If the dissolved oxygen level is the dominant control on the iodate concentration, then I/Ca can, to a first-order approximation, be used as a proxy for dissolved oxygen.

The advantages of I/Ca in calcite as a proxy are several-fold. It is a relatively simple, fast, and cost-effective analysis on the inductively coupled plasma-mass spectrometry (ICP-MS), allowing the possibility to construct records at high temporal resolution. Small volumes of sample materials are required for analyses, allowing measurements on micro-drilled samples and microfossils such as foraminifera (<0.5 mg). I/Ca should record local redox conditions, and therefore can be used as a complementary tool and combined with other proxies reflecting global changes. Multiple I/Ca records for the same time interval from different locations and water depths (e.g., Lu et al., 2020; Zhou et al., 2015) can provide a unique opportunity to reconstruct ocean oxygenation truly in four dimensions (3D space and time). Since the conversion between iodide and iodate is very sensitive to low levels of dissolved oxygen (instead of euxinia, with H_2S present), I/Ca signals are expected to track hypoxia or suboxia, which cannot be reconstructed by most redox proxies commonly used in Earth System Science.

However, it is important to note the main limitations of I/Ca. I/Ca is certainly not immune to post-depositional alteration (Hardisty et al., 2017), just as any other carbonate-based proxy. With a rising interest in measuring I/Ca in different laboratories, it is important to maintain consistency in analytical methods. Owing to the local nature of the I/Ca signal, sampling density may impact the statistical distribution of values. The potentially slow kinetics of iodide reoxidation imply that low iodate –high O_2 waters may exist, highlighting the possibility of I/Ca signals influenced by both oxygenation and water mass mixing. Based on all the foregoing, a major pitfall in the use of I/Ca is applying this proxy prematurely for quantitative interpretation of past oxygen levels.

3 Methods

Iodine concentrations in carbonate were first measured through pyrohydrolysis followed by ICP-MS measurement of the trap solutions (Muramatsu and Wedepohl, 1998). That method was modified to measure I/Ca in carbonate rocks (Lu et al., 2010). In preparation, rock samples are crushed (or drilled) and homogenized to a fine powder. Then ~4 mg of powdered sample is weighed and thoroughly rinsed with 18 MΩ water to remove soluble iodine (adsorbed or residual salt). Diluted nitric acid (3%) is added to dissolve carbonate. The solutions are diluted to a consistent matrix with ~50 ppm Ca, and 5 ppb indium and cesium are used as internal standards. The final solutions contain 0.5% tertiary amine to stabilize iodine in solution and ensure a reasonably short rinse time on the ICP-MS. The measurements immediately follow solution preparation to minimize changes in iodine speciation or volatilization. Fresh calibration standards are made from KIO_3 powders daily, for each batch of measurements. In the past, nitric acid was added to the final solutions loaded into the nebulizer to make them mildly acidic, and iodine was measured with a range of trace elements such as Mn, Sr, and rare earth elements (Lu et al., 2010). However, this methodology has evolved to use of an alkaline matrix and measurement of iodine with Ca and Mg only, to better avoid iodine speciation change due to the addition of acids, while all other trace elements are measured in separate runs in acidic solutions (e.g., Lu et al., 2016).

I/Ca measurements can be performed on a sensitive quadrupole ICP-MS (e.g., Varian/Bruker models) or magnetic sector ICP-MS (Thermo Element Series). The sensitivity of I-127 is typically tuned to 80,000–100,000 counts per second for a 1 ppb standard at Syracuse University and formerly at the University of Oxford. It is important to report the sensitivity of the 1 ppb I-127 standard on different instruments, especially when attempting to resolve I/Ca signals of 0–1 µmol/mol or even 0–0.5 µmol/mol. Instruments with lower sensitivity can produce useful data in some cases, such as large stratigraphic variations in I/Ca. The reliability of I/Ca data in a particular study should not be judged solely by the instrument sensitivity, but should be evaluated in the broader context, including stratigraphic settings and multiproxy comparison. However, the instrumental sensitivity on I-127 should be reported as metadata, so that the community can gradually develop a sense for which instruments are suitable for which types of I/Ca applications. Furthermore, I-127 concentrations in the solutions fed into the ICP-MS should be monitored over time. For example, groundwater and marine porewater may contain high levels of iodine and raise instrumental background, making it difficult for carbonate I/Ca measurements.

JCp-1 (coral, aragonite) is a calcium carbonate reference material with a published iodine concentration (Chai and Muramatsu, 2007), suitable for

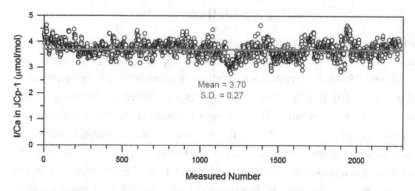

Figure 1 I/Ca values measured in reference material JCp-1 at Syracuse University laboratory in 2011–2018 (n = 2,280).

quality control and interlaboratory comparison. Earlier measurements of JCp-1 resulted in I/Ca values of 4.33 ± 0.36 μmol/mol (n = 5) (Gakushuin University, Japan) (Chai and Muramatsu, 2007) and 4.27 ± 0.06 μmol/mol (n = 8) (University of Oxford, UK) (Lu et al., 2010). More recent work reported lower values of 3.82 ± 0.39 μmol/mol (n = 60) (GEOMAR, Kiel, Germany) (Glock et al., 2014), with similar values reported in Hardisty et al. (2017) (UC Riverside, USA). The long-term (2012–2018) records of JCp-1 at the Syracuse University laboratory yielded an average I/Ca value of 3.70 ± 0.27 (1 standard deviation) μmol/mol (n = 2,280) (Fig. 1). A JCp-1 measurement is now run between every three to six unknown samples at the Syracuse laboratory. To ensure long-term consistency, the I/Ca values of unknown samples are corrected by adjusting the value of the adjacent JCp-1 measurements to 4.27.

Unfortunately, JCp-1 has been banned from export from Japan in recent years, so an in-house standard (synthetic calcite) has been produced and calibrated to JCp-1. The protocol of preparing synthetic calcite was described in the supplementary materials in Lu et al. (2010). Existing batches of synthetic calcite showed I/Ca values of 3.52 ± 0.27 μmol/mol (n = 248) after corrections with the JCp-1 value measured immediately before the in-house standard. After the I/Ca value has become consistently reproducible in different batches of this in-house standard over a few years, it will be made available to other laboratories on request.

Foraminiferal tests are commonly used in paleoceanographic applications, including isotope and trace element analyses. The cleaning procedure for monospecific foraminiferal tests for I/Ca analyses are modified after Barker et al. (2003) and Lu et al. (2010) and include the removal of clays by rinsing ultrasonically with deionized water, and removal of organic matter by oxidative cleaning with NaOH–H_2O_2 solutions (Glock et al., 2014; Hoogakker et al.,

2018; Lu et al., 2016). In addition to typical ICP-MS measurements, foraminiferal I/Ca measurements on individual specimens have been attempted with secondary ion mass spectrometry (SIMS) (Glock et al., 2016). High-resolution SIMS-derived I/Ca values within individual specimens (inter- and intratest) may provide information on short-term oxygen variability, whereas bulk ICP-MS analyses on whole specimens may be more suitable for relatively longer term oxygen reconstructions (Glock et al., 2016).

4 Sample Materials and Post-depositional Alterations

Criteria for selecting marine carbonate materials for I/Ca analysis depend on the specific application or research question. For early (e.g., Precambrian) Earth studies, relatively well-preserved dolomites or limestones are suitable for bulk carbonate I/Ca measurements. For samples of younger ages, bivalve shells should be avoided when measuring bulk carbonate I/Ca, since they seem to exclude iodine completely during calcification. Corals and foraminifera generally contain measurable iodine.

A large set of Neogene–Quaternary bulk carbonates (rather than specific fossils) with a varying degree of preservation was used to study the impact of diagenetic processes on the I/Ca signal (Hardisty et al., 2017). Carbonate recrystallization during subaerial exposure to meteoric waters, marine burial in anoxic pore waters, and dolomitization are most likely to reduce carbonate-associated iodine concentrations, likely reflecting the tendency of iodate to be reduced in pore waters. No post-depositional processes tested so far increase the I/Ca in carbonates. Thus, carbonate I/Ca values of ancient rocks provide minimum estimates of local seawater iodate levels (Hardisty et al., 2017). I/Ca analysis can be paired with typical diagenetic screening, such as Mg, Mn, δ^{13} C, δ^{18}O, and cathodoluminescence (e.g., Edwards et al., 2018; Lu et al., 2017; Wei et al., 2019; Wörndle et al., 2019). It is unlikely that any diagenetic proxy will be able to quantitatively and reliably resolve the primary signal of I/Ca. Like any other paleo-environmental proxy, I/Ca should be interpreted carefully, with multiproxy comparison and data-model comparison, and evaluation of the status of carbonate preservation.

5 Long-Term I/Ca Trends Through Earth History

5.1 I/Ca Excursions and pO_2 Rise

I/Ca has been applied at different timescales throughout Earth history, with the long-term record showing both excursions and step changes. I/Ca peaks were recorded in marine carbonates at almost all previously documented major instances of atmospheric O_2 rise, whether well-established or debated (Fig. 2).

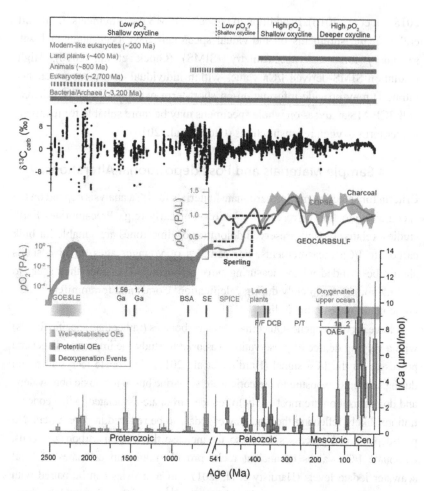

Figure 2 Summary of the long-term carbonate I/Ca record, modified from Lu et al. (2018), and updated with sections from Lowery et al. (2018), Shang et al. (2019), Wei et al. (2019), and Young et al. (2019). The eukaryotic evolution is modified from Knoll (2014). The $\delta^{13}C$ record is from Saltzman and Thomas (2012). The "modern-like eukaryotes" include both "calcifying eukaryote marine plankton" and "the modern evolutionary fauna" in marine vertebrates and invertebrates in the sense of Sepkoski (1981).

No I/Ca signal was detected in the Archean, whereas I/Ca signals can be measured for times during and after the Great Oxidation Event (GOE) at ~2.4 Ga (Hardisty et al., 2014).

Measurement of I/Ca across the GOE is a straightforward application of this proxy to detect the presence of dissolved oxygen during the first rise of O_2 in the atmosphere. The Lomagundi Event at ~2.1 Ga, sometimes considered to be part of the GOE, was suggested to be an "overshoot" of pO_2 (Bekker and Holland, 2012). The I/Ca values then declined in post-GOE Paleoproterozoic and Mesoproterozoic carbonates (~2–1 Ga), with notable exceptions at ~1.4 Ga (Hardisty et al., 2017) and ~1.56 Ga (Shang et al., 2019). Emerging evidence is building for a transient oxygenation event at ~1.4 Ga, for example, from molybdenum isotopes (Diamond et al., 2018) and uranium isotopes (Yang et al., 2017). Several I/Ca peaks in the Neoproterozoic and early Paleozoic (~800–400 Ma) provide evidence for oxygenation events, generally supported by independent geochemical proxies: the Bitter Springs carbon isotope anomaly at ~ 800 Ma (Isson et al., 2018; Wei et al., 2018b), the Ediacaran Shuram carbon isotope excursion at ~ 580 Ma (Sahoo et al., 2016; Shi et al., 2018; Tostevin et al., 2019; Wei et al., 2018a), and the rise of land plants at ~400 Ma (Dahl et al., 2010; He et al., 2019; Wallace et al., 2017).

Relatively high oxygen regions in the early oceans might have been present in a patchy oasis-like fashion (Olson et al., 2013). In addition, a post-depositional decrease in signals would make high I/Ca values harder to find in ancient carbonates. Thus, when maximum values above a low background are used to argue for an oxygenation event, there commonly is the potential caveat associated with sampling density in time and space in these relatively deep time studies, and it is not surprising that higher maximum I/Ca values appear to be found when more sections and more samples per section are measured (Fig. 3).

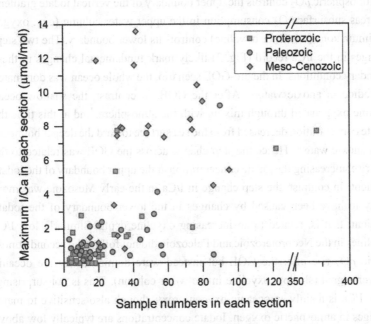

Figure 3 Sampling density vs. maximum I/Ca in each section (data shown in Fig. 2).

However, such cross-plots may exaggerate the sampling bias, particularly in Proterozoic studies. It is a common practice to measure more samples on parts of sections showing high values during preliminary studies (or across a well-known "event"), to obtain better stratigraphic constraints, when the target is high I/Ca values as an indicator of higher oxygen levels. To evaluate whether the maximum I/Ca values differentiating two time intervals were influenced by sampling density, total numbers of sections for each time interval should be compared, as well as total numbers of samples and time resolution.

5.2 Step Changes in I/Ca Baseline

In addition to the high values commonly associated with a global pO_2 increase, background I/Ca values show significant stepwise increases, first across the GOE and then in the Mesozoic (Fig. 2). Increasing diagenetic alteration with increasing age is predicted to cause gradual decreases in I/Ca, and changes in lithology (e.g., dolostone–limestone) do not correlate with the step changes observed here (Hardisty et al., 2017; Lu et al., 2018). Carbonate rocks selected for paleo-environmental reconstructions were precipitated primarily from/in the upper water column; thus the bulk carbonate I/Ca signal is largely controlled by the local upper ocean iodate gradient. I/Ca values are not expected to always correlate with redox proxies responding oxygenated seafloor globally.

Atmospheric pO_2 controls the upper boundary of the vertical iodate gradient, whereas subsurface O_2 consumption in the upper water column (e.g., oxygen minimum zone [OMZ] or oxycline) controls its lower boundary. The two step-changes in the I/Ca record (Fig. 2) likely mark fundamental changes in these boundary conditions. In the pre-GOE scenario, the whole ocean was dominated by iodide in anoxic waters. After the GOE, in contrast, the shallow ocean became oxygenated through mixing with the atmosphere, and at this time the iodate concentration decreased from the sea surface toward the deeper boundary with anoxic waters. Hence the step change across the GOE was related to the pO_2 rise, increasing the iodate concentration at the upper boundary of the iodate gradient. In contrast, the step change in I/Ca in the early Mesozoic was more likely to have been caused by changes in the lower boundary of the iodate gradient, that is, related to an increasing oxycline depth. Similarly low I/Ca baselines in the Neoproterozoic and Paleozoic did not follow the second atmospheric pO_2 rise after the GOE but instead were dominated by the oceanic oxygen signal (shallow oxycline in the water column). This is not surprising, since I/Ca is mainly an oceanic oxygen proxy, though also sensitive to major changes in atmospheric oxygen. Iodate concentrations are typically low above marine OMZs even under modern high atmospheric oxygen condition, implying that high pO_2 should not be automatically assumed to produce high I/ Ca baseline in long-term geologic records.

Oxycline depth is controlled by temperature-dependent solubility, ocean ventilation, and organic matter remineralization (Keeling et al., 2009). In the modern ocean, the surface water IO_3^- concentration can be low locally, during shallow hypoxic events (Truesdale and Bailey, 2000). Thus, high I/Ca measured in upper ocean carbonates requires both high pO_2 and a deep oxycline. Among the three factors (temperature-dependent solubility, ocean ventilation, and organic matter remineralization), the only one likely to behave in a stepwise manner on long timescales is the remineralization of organic matter, driven by biological evolution as well as environmental conditions. The influence of remineralization depth on oxycline and OMZ positions was modeled (Meyer et al., 2016). There was no proxy record demonstrating the timing of OMZ changes in the Phanerozoic, whereas the timing for proliferation of calcifying plankton is relatively well known (Fig. 2). The Phanerozoic I/Ca record tentatively points to the early Mesozoic for oxycline changes, consistent in timing with the diversification of calcifying plankton, although more work is needed to constrain the exact timing and rate of change (Lu et al., 2018).

6 Short-Term Ocean Deoxygenation Events

6.1 Paleozoic Extinctions

The Cambrian–Ordovician interval is known for several trilobite extinction events that have been linked to ocean anoxia (Saltzman et al., 2015). Two out of three sections in the Great Basin region, western United States, recorded lower I/Ca values coincident with a 30% extinction of standing generic diversity (Edwards et al., 2018). Coeval δ^{13} C and δ^{34}S excursions have been used to argue for increased global carbon burial and anoxia expansion. The same combination of proxy records (δ^{13} C, δ^{34}S, I/Ca) links Silurian events to ocean anoxia (Young et al., 2019 and Bowman et al., 2020). During the end-Devonian mass extinction (~359 Ma), low I/Ca marked an interval of deoxygenation more expanded than the interval of Hangenberg black shale deposition. The deoxygenation indicated by I/Ca is overall coeval with time intervals of carbon isotope excursion and biotic crisis indicated by fossil evidence (Liu et al., 2020).

The Permo-Triassic (P-T) mass extinction (~252 Ma) is the most severe biotic crisis in the Phanerozoic, during which possibly ~90% of marine species went extinct (Payne et al., 2004). Many lines of evidence, based on biotic proxies as well as organic and inorganic chemical proxies, document pervasive and widespread anoxia and even euxinia during this extinction event (e.g.,

Algeo et al., 2015; Brennecka et al., 2011; Elrick et al., 2017; Lau et al., 2016; Zhang et al., 2018). An early attempt to use carbonate iodine content as a redox proxy found very low I/Ca values across the P-T boundary, but concluded that these could be attributed to diagenesis (Loope et al., 2013).

6.2 Mesozoic Oceanic Anoxic Events

As a proof-of-concept, the iprOxy was first tested across Mesozoic Oceanic Anoxic Events (OAEs), generally associated with rapid warming and changes in the nutrient cycle, leading to large-scale, low-oxygen conditions in the oceans (Jenkyns, 2010). One iconic signature of OAEs is the positive carbon isotope excursions (CIE), reflecting enhanced global burial of organic matter. Records of the Toarcian OAE (~182 Ma) and the Cenomanian–Turonian OAE 2 (~92 Ma) generally show lower I/Ca during CIEs, as expected during the buildup of widespread reducing conditions in the oceans (Lu et al., 2010). In a follow-up study, carbonate I/Ca records across OAE 2 were obtained from seven sections and compared with Earth system modeling results (Zhou et al., 2015). Upper ocean oxygenation changes were spatially variable, and locally, some deoxygenation developed before the global CIE (Zhou et al., 2015), a scenario supported by trace metal and isotopic data (Owens et al., 2017; Them et al., 2018). I/Ca responded to a brief episode of reoxygenation due to global cooling during the Plenus cold event in the first half of OAE 2 (Zhou et al., 2015).

The I/TOC (total organic carbon) ratios in modern surface-subsurface sediments correlate with bottom water oxygen levels (Kennedy and Elderfield, 1987a, b; Lu et al., 2008; Price and Calvert, 1973). I/TOC records of the OAE 2 consistently show expansion of bottom-water deoxygenation during the CIE and a latitudinal gradient in agreement with ventilation patterns in an Earth system model (Zhou et al., 2017). The sedimentary iodine sink in the global ocean appears to be dependent on the flux of organic matter burial in oxygenation bottom water, since iodate (the oxidized form) is much more efficiently adsorbed onto organic matter than iodide (Zhou et al., 2017). Thus, the degree of global iodine drawdown during OAEs may be limited by iodate reduction in anoxic water lowering the efficiency of iodine adsorption onto organic materials.

OAEs were not generally associated with mass extinctions, possibly due to the presence of well-oxygenated ocean refuges, which may have existed, as suggested by I/Ca data for OAE 2. Alternatively, the seafloor area under anoxic waters may have been smaller than commonly considered (Clarkson et al., 2018). Another possibility is that Mesozoic faunas had greater physiological tolerance to warming and low-oxygen conditions, because they were adapted to a greenhouse world (Penn et al., 2018).

6.3 The Cretaceous–Paleogene Mass Extinction and Paleocene–Eocene Thermal Maximum

The Cretaceous–Paleogene (K–Pg) mass extinction (~66 Ma) destroyed ~76% of species on Earth, and was probably mainly caused by the impact of an asteroid (Schulte et al., 2010). There have been arguments that the K–Pg impact was followed by low-oxygen conditions; evidence for hypoxia appears to be local-regional, for example, at Caravaca and Agost (Spain) and Stevns Klint (Denmark) (Alegret et al., 2003; Vellekoop et al., 2018). Within the Chicxulub impact crater, well-oxygenated conditions existed during the recovery of the ecosystem after the asteroid impact, as suggested by bulk carbonate I/Ca values (Lowery et al., 2018). However, I/Ca data could not be reported for older intervals in that core at the impact site, in order to estimate oxygen levels prior to the event. There is thus as yet no evidence for global oceanic anoxia following the K–Pg impact, and bottom waters were not globally oxygen depleted, as shown by the nonextinction of benthic foraminifera (e.g., Alegret et al., 2003; Thomas, 1990).

The Paleocene–Eocene Thermal Maximum (PETM) (~55 Ma) was a rapid global warming event with a 5–8°C global temperature rise, associated with a major extinction of benthic foraminifera (McInerney and Wing, 2011). Globally low oceanic oxygen levels at this time were suggested by S-isotope (Yao et al., 2018) and Mo-isotope data (Dickson et al., 2012), and regional anoxia may have occurred in large epicontinental basins, for example, in the northern and eastern Tethys (Dickson et al., 2014). I/Ca in the bulk coarse fraction and planktic foraminifera provides evidence for upper ocean deoxygenation during the PETM, showing interbasinal oxygen gradients consistent with modeled oxycline depths (Zhou et al., 2014). I/Ca in deep-sea open ocean benthic foraminifera qualitatively indicates that bottom water oxygen levels were lower at intermediate depths during the late Paleocene–early Eocene (Zhou et al., 2016), broadly supporting redox-sensitive metal data at the same sites (Chun et al., 2010; Pälike et al., 2014).

7 Quaternary Glacial–Interglacial Cycles

Reconstructing glacial–interglacial changes in oceanic oxygen is important both for understanding the impact of changes in the carbon cycle on past climate conditions and for improving Earth system models that are commonly used to forecast future ocean deoxygenation. Glacial deep waters have been argued to have contained significantly lower mean O_2 because of increased oceanic carbon storage, possibly on the order of 10% (Broecker and Peng, 1982). However, establishing proxies to measure these relatively subtle changes in dissolved O_2 reliably is still challenging today.

Planktic foraminifera from Holocene core-top sediments contain low I/Ca (<2.5 µmol/mol) near OMZs and high I/Ca (e.g., >4 µmol/mol) in well-oxygenated locations (Lu et al., 2016, 2019). These core-top data further indicate that empirically planktic I/Ca is most sensitive to the hypoxic O_2 threshold (~70–100 µmol/kg), distinguished from $\delta^{15}N$, sensitive to the suboxic O_2 threshold (~5–10 µmol/kg) (Lu et al., 2019). However, researchers should use caution when interpreting planktic I/Ca as a proxy for *in situ* O_2 conditions strictly at the foraminiferal habitat. High O_2–low I/Ca or low IO_3^- conditions occur in the Southern Benguela region, with O_2 conditions highly variable on temporal and spatial scales (Lu et al., 2019). Low foraminiferal I/Ca values in this area may reflect short-lived low O_2 conditions, and/or the diffusion/advection of low-O_2 signal from nearby locations. A down-core planktic I/Ca record from the Pacific sector of the Southern Ocean indicates O_2-depleted waters close to planktic foraminiferal habitats during glacial periods, in contrast to the well-oxygenated interglacial periods (Lu et al., 2016). An I/Ca record from the eastern tropical Pacific relatively close to OMZs exhibited persistently low I/Ca values during the last 40 ka (Hoogakker et al., 2018).

Modern living benthic foraminifera from Peruvian OMZs show a positive relationship between I/Ca and a narrow range of ambient bottom water O_2 concentrations (2–35 µmol/kg) (Glock et al., 2014). I/Ca in epifaunal species (living above the water–sediment interface) are more likely to record bottom water O_2 signals, whereas I/Ca in infaunal species (living within the sediments, with the possibility of migrating vertically) are affected by pore water O_2 changes and thus may not well represent bottom water conditions. Data on epifaunal *Cibicidoides* spp. from global oceans show that epifaunal I/Ca does not linearly correlate with bottom water O_2, but decreases rapidly below an O_2 threshold (Lu et al., 2020). This threshold has not been precisely constrained, and a multiproxy approach leveraging additional independent oxygenation proxies [e.g., surface porosity of the foraminiferal test (Rathburn et al., 2018); $\Delta\delta^{13}C$ (Hoogakker et al., 2015)] will likely provide more reliable (semi-)quantitative estimates bottom water O_2 in down-core studies (Lu et al., 2020). I/Ca values in multiple benthic species in a core in the northeastern Pacific Ocean (955 m water depth) suggest that the bottom waters were well oxygenated during the Last Glacial Maximum (Taylor et al., 2017). A global compilation of five down-core I/Ca records and published (semi-)quantitative oxygenation proxy records indicates that low-O_2 waters (<50 µmol/kg) may have been more extensive in the glacial Atlantic and Pacific Oceans than today (Lu et al., 2020). The application of I/Ca in benthic foraminifera thus is promising, but as of today cannot be used to derive fully quantitative estimates of bottom water oxygen and requires more calibration and proxy development.

8 Future Prospects

There are many potential future applications and developments of I/Ca that can be targeted by different research communities. High Precambrian I/Ca values likely will continue to provide evidence for brief intervals of oxygenation (at least locally-regionally) against a generally low-oxygen background, contributing to the discussions about the dynamics of the transition from an anaerobic to aerobic world. The Paleozoic may be the period of final transition between low pO_2 and high pO_2 stasis, as argued by studies of carbon cycle perturbations (Bachan et al., 2017; Saltzman, 2005). Paleozoic oceans have traditionally been considered as well oxygenated, largely because of the abundant presence of eukaryote life forms including animals (Berner, 2006; Holland, 2006). More recent estimates of Paleozoic pO_2 levels, however, are notably different from the results of the GEOCARB family of models (Lenton et al., 2018; Sperling et al., 2015). pO_2 rise and ocean anoxia appear to be intertwined during the Paleozoic, and the relationship between them may need to be better illustrated. I/Ca may be a particularly useful tool to address this question, especially during the Devonian expansion of land plants, and marine extinction events possibly caused by anoxia (e.g., Frasnian–Famennian, end of Devonian).

The seawater iodine budget likely changed on tectonic timescales (Zhou et al., 2016), being controlled by the balance between sources and sinks that need to be better constrained (Lu et al., 2010). If increased organic matter burial and expanded bottom water anoxia forms a negative feedback preventing rapid drawdown of iodine during OAEs (Zhou et al., 2015), then the global iodine budget during oxygenation events becomes an intriguing object of study.

Stratigraphic records featuring decreases in I/Ca from high background values probably will reveal more nuance about ocean deoxygenation in a high pO_2 world. Increasing numbers of proxy data show deoxygenation both during warm greenhouse time periods (OAEs) and during cold glacial periods of the recent past and deep time (e.g., Bartlett et al., 2018; Hoogakker et al., 2018; Song et al., 2017). The temperature-controlled gas solubility thus does not by itself seem to dominate the dissolved O_2 levels in the ocean interior. More future research is needed to address the critical roles of ocean ventilation and the strength of the biological pump in determining ocean oxygenation under contrasting climatic conditions across geological timescales. Earth system models will likely form part of the solution to these questions.

Most Cenozoic intervals remain unexplored with regard to I/Ca. Recent studies suggest that large redox changes might have occurred in the Paleogene, including the potential presence of sulfidic conditions during the PETM (Yao et al., 2018), and widespread water column denitrification in the early Cenozoic (Kast et al., 2019). To compare I/Ca data with these redox proxies (e.g., S, N), the redox window to which foraminiferal I/Ca is sensitive needs to be calibrated. Single genus/species foraminiferal data from individual sites are needed, combined with additional information (e.g., on paleogeography, circulation patterns, paleo-productivity), in order to understand the dynamics of Cenozoic trends.

Foraminiferal I/Ca has great potential of serving as a semiquantitative proxy: I/Ca values below a certain threshold can be linked to dissolved O_2 values. It will likely be difficult to calibrate I/Ca to a numerical expression for a continuous range of O_2 values, as has been done for many fully quantitative proxies (e.g., Mg/Ca; Elderfield and Ganssen, 2000). The use of the quantitative $\Delta\delta^{13}C$ proxy for dissolved O_2 (the $\delta^{13}C$ difference between epifaunal benthic foraminiferal *Cibicidoides wuellerstorfi* and infaunal *Globobulimina* spp.), as an example, is limited in its application to the occurrence of particular species of foraminifera, which are available in limited locations and relatively recent time intervals only (Hoogakker et al., 2015). Thus, a semiquantitative iprOxy with versatile applications likely will assist reconstructing the coevolution of biosphere and geosphere, and contribute useful and unique information to Earth System Science.

References

Alegret, L., Molina, E., and Thomas, E. (2003) Benthic foraminiferal turnover across the Cretaceous/Paleogene boundary at Agost (southeastern Spain): Paleoenvironmental inferences. *Marine Micropaleontology* 48: 251–279.

Algeo, T. J., Chen, Z.-Q., and Bottjer, D. J. (2015) Global review of the Permian–Triassic mass extinction and subsequent recovery: Part II. *Earth-Science Reviews* 100: 1–4.

Amachi, S., Kawaguchi, N., Muramatsu, Y., Tsuchiya, S., Watanabe, Y., Shinoyama, H., and Fujii, T. (2007) Dissimilatory iodate reduction by marine *Pseudomonas* sp. strain SCT. *Applied and Environmental Microbiology* 73: 5725–5730.

Bachan, A., Lau, K. V., Saltzman, M. R., Thomas, E., Kump, L. R., and Payne, J. L. (2017) A model for the decrease in amplitude of carbon isotope excursions across the Phanerozoic. *American Journal of Science* 317: 641–676.

Barker, S., Greaves, M., and Elderfield, H. (2003) A study of cleaning procedures used for foraminiferal Mg/Ca paleothermometry. *Geochemistry, Geophysics, Geosystems* 4: 8407.

Bartlett, R., Elrick, M., Wheeley, J. R., Polyak, V., Desrochers, A., and Asmerom, Y. (2018) Abrupt global-ocean anoxia during the Late Ordovician–early Silurian detected using uranium isotopes of marine carbonates. *Proceedings of the National Academy of Sciences of the USA* 115: 5896–5901.

Bekker, A., and Holland, H. (2012) Oxygen overshoot and recovery during the early Paleoproterozoic. *Earth and Planetary Science Letters* 317: 295–304.

Berner, R. A. (2006) GEOCARBSULF: A combined model for Phanerozoic atmospheric O_2 and CO_2. *Geochimica et Cosmochimica Acta* 70: 5653–5664.

Bowman, C. N., Lindskog, A., Kozik, N. P., Richbourg, C. G., Owens, J. D., & Young, S. A. (2020). Integrated sedimentary, biotic, and paleoredox dynamics from multiple localities in southern Laurentia during the late Silurian (Ludfordian) extinction event. Palaeogeography, Palaeoclimatology, Palaeoecology 553, 109799.

Brennecka, G. A., Herrmann, A. D., Algeo, T. J., and Anbar, A. D. (2011) Rapid expansion of oceanic anoxia immediately before the end-Permian mass extinction. *Proceedings of the National Academy of Sciences of the USA* 108: 17631–17634.

Broecker, W., and Peng, T. (1982) *Tracers in the sea*. Palisades, NY: Lamont-Doherty Geological Observatory.

Chai, J. Y., and Muramatsu, Y. (2007) Determination of bromine and iodine in twenty-three geochemical reference materials by ICP-MS. *Geostandards and Geoanalytical Research* 31: 143–150.

Chance, R., Baker, A. R., Carpenter, L., and Jickells, T. D. (2014) The distribution of iodide at the sea surface. *Environmental Science: Processes & Impacts* 16: 1841–1859.

Chun, C. O., Delaney, M. L., and Zachos, J. C. (2010) Paleoredox changes across the Paleocene-Eocene thermal maximum, Walvis Ridge (ODP Sites 1262, 1263, and 1266): Evidence from Mn and U enrichment factors. *Paleoceanography* 25: PA4202.

Clarkson, M. O., Stirling, C. H., Jenkyns, H. C., et al. (2018) Uranium isotope evidence for two episodes of deoxygenation during Oceanic Anoxic Event 2. *Proceedings of the National Academy of Sciences of the USA* 115: 2918–2923.

Cutter, G. A., Moffett, J. W., Nielsdóttir, M. C., and Sanial, V. (2018) Multiple oxidation state trace elements in suboxic waters off Peru: In situ redox processes and advective/diffusive horizontal transport. *Marine Chemistry* 201: 77–89.

Dahl, T. W., Hammarlund, E. U., Anbar, A. D., et al. (2010) Devonian rise in atmospheric oxygen correlated to the radiations of terrestrial plants and large predatory fish. *Proceedings of the National Academy of Sciences of the USA* 107: 17911–17915.

Diamond, C., Planavsky, N., Wang, C., and Lyons, T. (2018) What the~ 1.4 Ga Xiamaling Formation can and cannot tell us about the mid-Proterozoic ocean. *Geobiology* 16: 219–236.

Dickson, A. J., Cohen, A. S., and Coe, A. L. (2012) Seawater oxygenation during the Paleocene-Eocene thermal maximum. *Geology* 40: 639–642.

Dickson, A. J., Rees-Owen, R. L., März, C., et al. (2014) The spread of marine anoxia on the northern Tethys margin during the Paleocene-Eocene Thermal Maximum. *Paleoceanography* 29: 471–488.

Edwards, C. T., Fike, D. A., Saltzman, M. R., Lu, W., and Lu, Z. (2018) Evidence for local and global redox conditions at an Early Ordovician (Tremadocian) mass extinction. *Earth and Planetary Science Letters* 481: 125–135.

Elderfield, H., and Ganssen, G. (2000) Past temperature and $\delta\ ^{18}O$ of surface ocean waters inferred from foraminiferal Mg/Ca ratios. *Nature* 405: 442–445.

Elrick, M., Polyak, V., Algeo, T. J., et al. (2017) Global-ocean redox variation during the middle-late Permian through Early Triassic based on uranium isotope and Th/U trends of marine carbonates. *Geology* 45: 163–166.

Farrenkopf, A. M., Dollhopf, M. E., Chadhain, S. N., Luther, G. W., and Nealson, K. H. (1997a) Reduction of iodate in seawater during Arabian Sea shipboard incubations and in laboratory cultures of the marine bacterium *Shewanella putrefaciens* strain MR-4. *Marine Chemistry* 57: 347–354.

Farrenkopf, A. M., and Luther, G. W. (2002) Iodine chemistry reflects productivity and denitrification in the Arabian Sea: Evidence for flux of dissolved species from sediments of western India into the OMZ: Deep Sea Research Part II. *Topical Studies in Oceanography* 49: 2303–2318.

Farrenkopf, A. M., Luther, G. W., Truesdale, V. W., and Van der Weijden, C. H. (1997b) Sub-surface iodide maxima: Evidence for biologically catalyzed redox cycling in Arabian Sea OMZ during the SW intermonsoon: Deep Sea Research Part II. *Topical Studies in Oceanography* 44: 1391–1409.

Feng, X., and Redfern, S. A. (2018) Iodate in calcite, aragonite and vaterite $CaCO_3$: Insights from first-principles calculations and implications for the I/Ca geochemical proxy. *Geochimica et Cosmochimica Acta* 236: 351–360.

Glock, N., Liebetrau, V., and Eisenhauer, A. (2014) I/Ca ratios in benthic foraminifera from the Peruvian oxygen minimum zone: analytical methodology and evaluation as proxy for redox conditions. *Biogeosciences* 11: 7077–7095.

Glock, N., Liebetrau, V., Eisenhauer, A., and Rocholl, A. (2016) High resolution I/Ca ratios of benthic foraminifera from the Peruvian oxygen-minimum-zone: A SIMS derived assessment of a potential redox proxy. *Chemical Geology* 447: 40–53.

Hardisty, D. S., Horner, T. J., Wankel, S. D., Blusztajn, J., and Nielsen, S. G. (2020) Experimental observations of marine iodide oxidation using a novel sparge-interface MC-ICP-MS technique. *Chemical Geology* 532: 11936.

Hardisty, D. S., Lu, Z., Bekker, A., et al. (2017) Perspectives on Proterozoic surface ocean redox from iodine contents in ancient and recent carbonate. *Earth and Planetary Science Letters* 463: 159–170.

Hardisty, D. S., Lu, Z., Planavsky, N. J., et al. (2014) An iodine record of Paleoproterozoic surface ocean oxygenation. *Geology* 42: 619–622.

He, R., Lu, W., Junium, C. K., Ver Straeten, C. A., others Lu, Z. (2019). Paleo-redox context of the Mid-Devonian Appalachian Basin and its relevance to biocrises. *Geochimica et Cosmochimica Acta*, https://doi.org/10.1016/j.gca.2019.12.019

Holland, H. D. (2006) The oxygenation of the atmosphere and oceans. *Philosophical Transactions of the Royal Society B: Biological Sciences* 361: 903–915.

Hoogakker, B. A., Elderfield, H., Schmiedl, G., McCave, I. N., and Rickaby, R. E. (2015) Glacial-interglacial changes in bottom-water oxygen content on the Portuguese margin. *Nature Geoscience* 8: 40.

Hoogakker, B. A., Lu, Z., Umling, N., et al. (2018) Glacial expansion of oxygen-depleted seawater in the eastern tropical Pacific. *Nature* 562: 410.

Isson, T. T., Love, G. D., Dupont, C. L., et al. (2018) Tracking the rise of eukaryotes to ecological dominance with zinc isotopes. *Geobiology* 16: 341–352.

Jenkyns, H. C. (2010) Geochemistry of oceanic anoxic events. *Geochemistry, Geophysics, Geosystems* 11: Q03004.

Kast, E. R., Stolper, D. A., Auderset, A., et al. (2019) Nitrogen isotope evidence for expanded ocean suboxia in the early Cenozoic. *Science* 364: 386–389.

Keeling, R. F., Körtzinger, A., and Gruber, N. (2009) Ocean deoxygenation in a warming world. *Annual Review of Marine Science* 2: 463–493.

Kennedy, H., and Elderfield, H. (1987a) Iodine diagenesis in non-pelagic deep-sea sediments. *Geochimica et Cosmochimica Acta* 51: 2505–2514.

Kennedy, H., and Elderfield, H. (1987b) Iodine diagenesis in pelagic deep-sea sediments. *Geochimica et Cosmochimica Acta* 51: 2489–2504.

Kerisit, S. N., Smith, F. N., Saslow, S. A., Hoover, M. E., Lawter, A. R., and Qafoku, N. P. (2018) Incorporation modes of iodate in calcite. *Environmental Science & Technology* 52: 5902–5910.

Knoll, A. H. (2014) Paleobiological perspectives on early eukaryotic evolution. *Cold Spring Harbor Perspectives in Biology* 6: a016121.

Lau, K. V., Maher, K., Altiner, D., et al. (2016) Marine anoxia and delayed Earth system recovery after the end-Permian extinction. *Proceedings of the National Academy of Sciences of the USA* 113: 2360–2365.

Lenton, T. M., Daines, S. J., and Mills, B. J. (2018) COPSE reloaded: An improved model of biogeochemical cycling over Phanerozoic time. *Earth-Science Reviews* 178: 1–28.

Liu, J., Luo, G., Lu, Z., Lu, W., Qie, W., Zhang, F., Wang, X., & Xie, S. (2019). Intensified ocean deoxygenation during the end Devonian mass extinction. *Geochemistry, Geophysics, Geosystems*, 20(12), 6187–6198.

Loope, G. R., Kump, L. R., and Arthur, M. A. (2013) Shallow water redox conditions from the Permian–Triassic boundary microbialite: The rare earth element and iodine geochemistry of carbonates from Turkey and South China. *Chemical Geology* 351: 195–208.

Lowery, C. M., Bralower, T. J., Owens, J. D., et al. (2018) Rapid recovery of life at ground zero of the end-Cretaceous mass extinction. *Nature* 558: 288.

Lu, W., Dickson, A. J., Thomas, E., Rickaby, R. E., Chapman, P., & Lu, Z. (2019). Refining the planktic foraminiferal I/Ca proxy: Results from the Southeast Atlantic Ocean. *Geochimica et Cosmochimica Acta*, https://doi.org/10.1016/j.gca.2019.10.025

Lu, W., Rickaby, R., Hoogakker, B., et al. 2020 I/Ca in epifaunal benthic foraminifera: A semi-quantitative proxy for bottom water oxygen in a multi-proxy compilation for glacial ocean deoxygenation. *Earth and Planetary Science Letters* 533: 116055.

Lu, W., Ridgwell, A., Thomas, E., et al. (2018) Late inception of a resiliently oxygenated upper ocean. *Science* 361: 174–177.

Lu, W., Wörndle, S., Halverson, G., et al. (2017) Iodine proxy evidence for increased ocean oxygenation during the Bitter Springs Anomaly. *Geochemical Perspective Letters* 5: 53–57.

Lu, Z., Hensen, C., Fehn, U., and Wallmann, K. (2008) Halogen and 129I systematics in gas hydrate fields at the northern Cascadia margin (IODP Expedition 311): Insights from numerical modeling. *Geochemistry, Geophysics, Geosystems* 9: Q10006.

Lu, Z., Hoogakker, B. A., Hillenbrand, C.-D., et al. (2016) Oxygen depletion recorded in upper waters of the glacial Southern Ocean. *Nature Communications* 7: 11146.

Lu, Z., Jenkyns, H. C., and Rickaby, R. E. (2010) Iodine to calcium ratios in marine carbonate as a paleo-redox proxy during oceanic anoxic events. *Geology* 38: 1107–1110.

Luther, G. W., and Campbell, T. (1991) Iodine speciation in the water column of the Black Sea. Deep Sea Research Part A. Oceanographic Research Papers 38.

Luther, G. W., Wu, J., and Cullen, J. B. (1995) Redox chemistry of iodine in seawater: frontier molecular orbital theory considerations. *Advances in Chemistry* 244: 135.

Lyons, T. W., Reinhard, C. T., and Planavsky, N. J. (2014) The rise of oxygen in Earth's early ocean and atmosphere. *Nature* 506: 307–315.

McInerney, F. A., and Wing, S. L. (2011) The Paleocene-Eocene Thermal Maximum: A perturbation of carbon cycle, climate, and biosphere with implications for the future. *Annual Review of Earth and Planetary Sciences* 39: 489–516.

Meyer, K., Ridgwell, A., and Payne, J. (2016) The influence of the biological pump on ocean chemistry: Implications for long-term trends in marine redox chemistry, the global carbon cycle, and marine animal ecosystems. *Geobiology* 14: 207–219.

Muramatsu, Y., and Wedepohl, K. H. (1998) The distribution of iodine in the earth's crust. *Chemical Geology* 147: 201–216.

Olson, S. L., Kump, L. R., and Kasting, J. F. (2013) Quantifying the areal extent and dissolved oxygen concentrations of Archean oxygen oases. *Chemical Geology* 362: 35–43.

Owens, J. D., Lyons, T. W., Hardisty, D. S., Lowery, C. M., Lu, Z., Lee, B., and Jenkyns, H. C. (2017) Patterns of local and global redox variability during the Cenomanian–Turonian Boundary Event (Oceanic Anoxic Event 2) recorded in carbonates and shales from central Italy. *Sedimentology* 64: 168–185.

Pälike, C., Delaney, M. L., and Zachos, J. C. (2014) Deep-sea redox across the Paleocene-Eocene thermal maximum. *Geochemistry, Geophysics, Geosystems* 15: 1038–1053.

Payne, J. L., Lehrmann, D. J., Wei, J., Orchard, M. J., Schrag, D. P., and Knoll, A. H. (2004) Large perturbations of the carbon cycle during recovery from the end-Permian extinction. *Science* 305: 506–509.

Penn, J. L., Deutsch, C., Payne, J. L., and Sperling, E. A. (2018) Temperature-dependent hypoxia explains biogeography and severity of end-Permian marine mass extinction. *Science* 362: eaat1327.

Podder, J., Lin, J., Sun, W., et al. (2017) Iodate in calcite and vaterite: Insights from synchrotron X-ray absorption spectroscopy and first-principles calculations. *Geochimica et Cosmochimica Acta* 198: 218–228.

Price, N., and Calvert, S. (1973) The geochemistry of iodine in oxidised and reduced recent marine sediments. *Geochimica et Cosmochimica Acta* 37: 2149–2158.

Rathburn, A. E., Willingham, J., Ziebis, W., Burkett, A. M., and Corliss, B. H. (2018) A new biological proxy for deep-sea paleo-oxygen: Pores of epifaunal benthic foraminifera. *Scientific Reports* 8: Article 9456.

Rue, E. L., Smith, G. J., Cutter, G. A., and Bruland, K. W. (1997) The response of trace element redox couples to suboxic conditions in the water column. *Deep Sea Research Part I: Oceanographic Research Papers* 44: 113–134.

Sahoo, S. K., Planavsky, N., Jiang, G., et al. (2016) Oceanic oxygenation events in the anoxic Ediacaran ocean. *Geobiology* 14: 457–468.

Saltzman, M., and Thomas, E. (2012) Carbon isotope stratigraphy. In F. M. Gradstein, J. G. Ogg, M. B. Schmitz, and G. M. Ogg (eds.), *The geologic time scale*, Vol. 1 (pp. 207–232). Oxford: Elsevier BV.

Saltzman, M. R. (2005) Phosphorus, nitrogen, and the redox evolution of the Paleozoic oceans. *Geology* 33: 573–576.

Saltzman, M. R., Edwards, C. T., Adrain, J. M., and Westrop, S. R. (2015) Persistent oceanic anoxia and elevated extinction rates separate the Cambrian and Ordovician radiations. *Geology* 43: 807–810.

Schulte, P., Alegret, L., Arenillas, I., et al. (2010) The Chicxulub asteroid impact and mass extinction at the Cretaceous-Paleogene boundary. *Science* 327: 1214–1218.

Sepkoski, J. J. (1981) A factor analytic description of the Phanerozoic marine fossil record. *Paleobiology* 7: 36–53.

Shang, M., Tang, D., Shi, X., Zhou, L., Zhou, X., Song, H., and Jiang, G. (2019) A pulse of oxygen increase in the early Mesoproterozoic ocean at ca. 1.57–1.56 Ga. *Earth and Planetary Science Letters* 527: 115797.

Shi, W., Li, C., Luo, G., et al. (2018) Sulfur isotope evidence for transient marine-shelf oxidation during the Ediacaran Shuram Excursion. *Geology* 46: 267–270.

Song, H., Song, H., Algeo, T. J., et al. (2017) Uranium and carbon isotopes document global-ocean redoxproductivity relationships linked to cooling during the Frasnian-Famennian mass extinction. *Geology* 45: 887–890.

Sperling, E. A., Wolock, C. J., Morgan, A. S., et al. (2015) Statistical analysis of iron geochemical data suggests limited late Proterozoic oxygenation. *Nature* 523: 451.

Taylor, M., Hendy, I., and Chappaz, A. (2017) Assessing oxygen depletion in the Northeastern Pacific Ocean during the last deglaciation using I/Ca ratios from multiple benthic foraminiferal species. *Paleoceanography* 32: 746–762.

Them, T. R., Gill, B. C., Caruthers, A. H., et al. (2018) Thallium isotopes reveal protracted anoxia during the Toarcian (Early Jurassic) associated with volcanism, carbon burial, and mass extinction. *Proceedings of the National Academy of Sciences of the USA* 115: 6596–6601.

Thomas, E. (1990) Late Cretaceous–early Eocene mass extinctions in the deep sea. *Geological Society of America Special Publication* 247: 481–495.

Tostevin, R., Clarkson, M. O., Gangl, S., et al. (2019) Uranium isotope evidence for an expansion of anoxia in terminal Ediacaran oceans. *Earth and Planetary Science Letters* 506: 104–112.

Truesdale, V., and Bailey, G. (2000) Dissolved iodate and total iodine during an extreme hypoxic event in the Southern Benguela system. *Estuarine, Coastal and Shelf Science* 50: 751–760.

Vellekoop, J., Woelders, L., van Helmond, N. A., et al. (2018) Shelf hypoxia in response to global warming after the Cretaceous-Paleogene boundary impact. *Geology* 46: 683–686.

Wallace, M. W., Shuster, A., Greig, A., Planavsky, N. J., and Reed, C. P. (2017) Oxygenation history of the Neoproterozoic to early Phanerozoic and the rise of land plants. *Earth and Planetary Science Letters* 466: 12–19.

Wei, H., Wang, X., Shi, X., Jiang, G., Tang, D., Wang, L., and An, Z. (2019) Iodine content of the carbonates from the Doushantuo Formation and shallow ocean redox change on the Ediacaran Yangtze Platform, South China. *Precambrian Research* 322: 160–169.

Wei, W., Frei, R., Gilleaudeau, G. J., Li, D., Wei, G.-Y., Chen, X., and Ling, H.-F. (2018a) Oxygenation variations in the atmosphere and shallow seawaters of the Yangtze Platform during the Ediacaran Period: Clues from Cr-isotope and Ce-anomaly in carbonates. *Precambrian Research* 313: 78–90.

Wei, W., Frei, R., Klaebe, R., Li, D., Wei, G.-Y., and Ling, H.-F. (2018b) Redox condition in the Nanhua Basin during the waning of the Sturtian glaciation: A chromium-isotope perspective. *Precambrian Research* 319: 198–210.

Wong, G. T., and Brewer, P. G. (1977) The marine chemistry of iodine in anoxic basins. *Geochimica et Cosmochimica Acta* 41: 151–159.

Wong, G. T., and Cheng, X. H. (2008) Dissolved inorganic and organic iodine in the Chesapeake Bay and adjacent Atlantic waters: Speciation changes through an estuarine system. *Marine Chemistry* 111: 221–232.

Wong, G. T., Takayanagi, K., and Todd, J. F. (1985) Dissolved iodine in waters overlying and in the Orca Basin, Gulf of Mexico. *Marine Chemistry* 2: 177–183.

Wong, G. T., and Zhang, L. S. (2003) Seasonal variations in the speciation of dissolved iodine in the Chesapeake Bay. *Estuarine, Coastal and Shelf Science* 56: 1093–1106.

Wörndle, S., Crockford, P. W., Kunzmann, M., Bui, T. H., and Halverson, G. P. (2019) Linking the Bitter Springs carbon isotope anomaly and early Neoproterozoic oxygenation through I/[Ca+Mg] ratios. *Chemical Geology* 524: 119–135.

Yang, S., Kendall, B., Lu, X., Zhang, F., and Zheng, W. (2017) Uranium isotope compositions of mid-Proterozoic black shales: Evidence for an episode of increased ocean oxygenation at 1.36 Ga and evaluation of the effect of post-depositional hydrothermal fluid flo. *Precambrian Research* 298: 187–201.

Yao, W., Paytan, A., and Wortmann, U. G. (2018) Large-scale ocean deoxygenation during the Paleocene-Eocene Thermal Maximum. *Science* 361: 804–806.

Young, S. A., Kleinberg, A., and Owens, J. (2019) Geochemical evidence for expansion of marine euxinia during an early Silurian (Llandovery–Wenlock boundary) mass extinction. *Earth and Planetary Science Letters* 513: 187–196.

Zhang, F., Romaniello, S. J., Algeo, T. J., et al. (2018) Multiple episodes of extensive marine anoxia linked to global warming and continental weathering following the latest Permian mass extinction. *Science Advances* 4: e1602921.

Zhou, X., Jenkyns, H. C., Lu, W., Hardisty, D. S., Owens, J. D., Lyons, T. W., and Lu, Z. (2017) Organically bound iodine as a bottom-water redox proxy: Preliminary validation and application. *Chemical Geology* 457: 95–106.

Zhou, X., Jenkyns, H. C., Owens, J. D., et al. (2015) Upper ocean oxygenation dynamics from I/Ca ratios during the Cenomanian-Turonian OAE 2. *Paleoceanography* 30: 510–526.

Zhou, X., Thomas, E., Rickaby, R. E., Winguth, A. M., and Lu, Z. (2014) I/Ca evidence for upper ocean deoxygenation during the PETM. *Paleoceanography* 29: 964–975.

Zhou, X., Thomas, E., Winguth, A., (2016) Expanded oxygen minimum zones during the late Paleocene-early Eocene: Hints from multiproxy comparison and ocean modeling. *Paleoceanography and Paleoclimatology* 31: 1532–1546.

Acknowledgments

We acknowledge funding from grants NSF EAR-1349252, OCE-1232620, and OCE-1736542 (Z. L.); NSF OCE-1736538 (E. T.); Wolfson Research Merit award from the Royal Society and ERC Consolidator grant no. 681746 (R. R.); and Syracuse University Graduate Fellowship and Research Excellence Doctoral Funding Fellowship (W. L.).

Cambridge Elements ≡

Elements in Geochemical Tracers in Earth System Science

Timothy Lyons
University of California

Timothy Lyons is a Distinguished Professor of Biogeochemistry in the Department of Earth Sciences at the University of California, Riverside. He is an expert in the use of geochemical tracers for applications in astrobiology, geobiology and Earth history. Professor Lyons leads the 'Alternative Earths' team of the NASA Astrobiology Institute and the Alternative Earths Astrobiology Center at UC Riverside.

Alexandra Turchyn
University of Cambridge

Alexandra Turchyn is a University Reader in Biogeochemistry in the Department of Earth Sciences at the University of Cambridge. Her primary research interests are in isotope geochemistry and the application of geochemistry to interrogate modern and past environments.

Chris Reinhard
Georgia Institute of Technology

Chris Reinhard is an Assistant Professor in the Department of Earth and Atmospheric Sciences at the Georgia Institute of Technology. His research focuses on biogeochemistry and paleoclimatology, and he is an Institutional PI on the 'Alternative Earths' team of the NASA Astrobiology Institute.

About the series

This innovative series provides authoritative, concise overviews of the many novel isotope and elemental systems that can be used as 'proxies' or 'geochemical tracers' to reconstruct past environments over thousands to millions to billions of years—from the evolving chemistry of the atmosphere and oceans to their cause-and-effect relationships with life.

Covering a wide variety of geochemical tracers, the series reviews each method in terms of the geochemical underpinnings, the promises and pitfalls, and the 'state-of-the-art' and future prospects, providing a dynamic reference resource for graduate students, researchers and scientists in geochemistry, astrobiology, paleontology, paleoceanography and paleoclimatology.

The short, timely, broadly accessible papers provide much-needed primers for a wide audience—highlighting the cutting-edge of both new and established proxies as applied to diverse questions about Earth system evolution over wide-ranging time scales.

Cambridge Elements ⁼

Elements in Geochemical Tracers in Earth System Science

Elements in the series